D0459454

ALEXANDER FLEMING

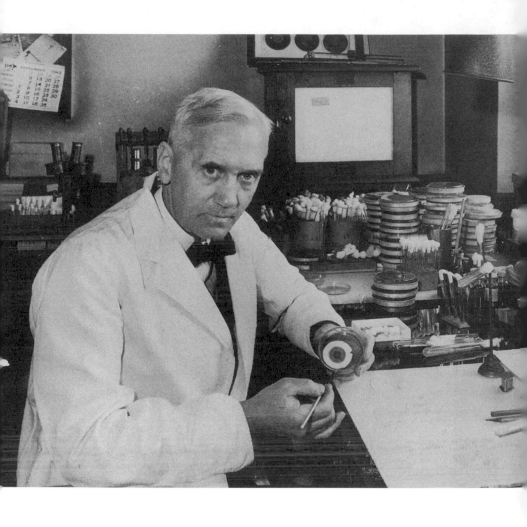

ALEXANDER FLEMING

Discoverer of Penicillin
by Ted Gottfried

A Book Report Biography
FRANKLIN WATTS
A Division of Grolier Publishing
New York London Hong Kong Sydney
Danbury, Connecticut

In memory of Dr. James Kaufman.
He was a physician who cared for his patients—and cared
about them.

frontispiece: *Alexander Fleming works at his desk at St. Mary's Hospital Medical School in 1943. Fleming's discovery of lysozyme and penicillin led to the development of many drugs that have saved countless lives.*

Cover illustration by Julian Allen

Photographs ©: AP/Wide World Photos: 87, 95; Archive Photos/Popperfoto: 99; Corbis-Bettmann: 13, 34, 44, 51, 83; Hulton Deutsch Collection Limited: 24, 50; Imperial College School of Medicine at St. Mary's, London: 31, 54, 60; Peter Arnold Inc.: 38, 63 (Manfred Kage); Photo Researchers: 42 (ASM/SS), 70 (Ken Edward/SS), 81 (SPL), 17 (St. Mary's Hospital Medical School/SPL); UPI/Corbis-Bettmann: 2, 89.

Library of Congress Cataloging-in-Publication Data

Alexander Fleming: discoverer of penicillin / Ted Gottfried.
    p.   cm.—(A book report biography)
Includes bibliographical references and index.
Summary: A biography of the British bacteriologist, born in Scotland, who was knighted and awarded the 1945 Nobel Prize in medicine for discovering penicillin.
ISBN 0-531-11370-1
1. Fleming, Alexander, 1881–1955—Juvenile literature. 2. Bacteriologists—Great Britain—Biography—Juvenile literature. 3. Penicillin—History—Juvenile literature. [1. Fleming, Alexander, 1881–1955. 2. Scientists. 3. Penicillin—History.] I. Title II. Series.
QR31. F5G68 1997
616'.014'092—dc21
[B]   97-7671
                                                                      CIP
                                                                      AC

© 1997 by Ted Gottfried
All rights reserved. Published simultaneously in Canada
Printed in the United States of America
1 2 3 4 5 6 7 8 9 10 R 06 05 04 03 02 01 00 99 98 97

CONTENTS

ACKNOWLEDGMENTS

I am indebted to the following people for their support on this book: Anita Yulsman, R. D. H., and George Fried for their technical assistance; author Janet Bode for her YA expertise; the personnel of various branches of the New York Public Library, and, as always, my librarian wife, Harriet Gottfried, who read and critiqued the manuscript in progress. Their help was invaluable, but any errors, or shortcomings, are mine alone.

CHAPTER ONE

MIRACLE OF THE MOLD

The new year of 1929 did not begin happily for Stuart Craddock, a doctor at St. Mary's Hospital in London, England. He was suffering from an infected sinus. His nose was swollen. His eyes watered. His head was stuffed. He was miserable. On January 9, Craddock's colleague, Dr. Alexander Fleming, treated him with a new substance that Fleming had been developing. Craddock's condition improved immediately. As if by a miracle, the sinus **infcction** was cured. The remedy was **penicillin.**

A CHANCE DISCOVERY

A year carlicr, Fleming had been studying a new **antiseptic** called mercuric chloride. He was skeptical about its usefulness, thinking there was not much chance of locating any general antiseptic that would effectively destroy disease-causing

bacteria in the bloodstream. The problem was that a substance strong enough to kill the bacteria that caused the most serious infections and diseases also damaged the cells of the patient, sometimes fatally. Despite these doubts, however, Fleming was studying the effects of mercuric chloride on various **streptococci** bacteria.

Bacteria in the streptococci family cause many diseases, including scarlet fever, tonsillitis, and arthritis. Fleming was trying to determine whether a weak solution of mercuric chloride might be developed that would not harm human cells but would weaken the streptococcus enough so that they could not fight off the bacteria-eating **phagocytes,** or cells that engulf bacteria. He prepared many **culture** dishes with streptococcus and various strengths of mercuric chloride and observed and recorded the results regularly.

At the same time, he was working on an article for *A System of Bacteriology,* a major text being prepared by the Medical Research Council. This task required Fleming to prepare culture dishes of **Staphylococci**. This type of bacteria causes boils, abscesses, carbuncles, and similar pus-producing infections.

One autumn day a colleague, Merlin Pryce, wandered into Fleming's laboratory. As the two men chatted, Fleming browsed through the culture dishes. "That's funny!" He cut Pryce short with the exclamation. Light blue **mold** was growing on one of the culture dishes. There was nothing unusual about that. Molds grow everywhere. A cheese left standing too long will grow a greenish mold, as will stale bread. Damp cellars and basements develop molds where the walls meet the foundation; fungus molds grow at the base of trees; and molds often grow on culture dishes that have at some point been exposed to the air.

What attracted Fleming's attention this time was that there were clearly defined areas around the mold where the staphylococci bacteria had been wiped out. The mold had killed the hardy **microbes,** or disease-causing bacteria. Fleming scraped a piece of the mold from the dish with a scalpel and put it into a jar of common laboratory fluid that the lab workers called "broth." He knew the mold would grow in this liquid and could be observed.

"Instead of casting out the contaminated culture," he remarked later, "I made some investigations."

"Instead of casting out the contaminated culture, I made some investigations."

Pryce did not think the mold was that important. He supposed it had accidentally produced an acid that destroyed the staphylococci. Still, he was polite. "That's how you discovered lysozyme," he remarked to Alec. Fleming had previously discovered **lysozyme,** a substance produced by the body that defends it against harmful bacteria. But lysozyme was an **enzyme** manufactured by the body, not a mold. Molds are **parasites** that grow in many places and feed on a variety of substances. What Fleming had discovered was that a particular mold fed on a particular type of sturdy and harmful bacteria.

Within days, the mold that Fleming had put in the broth turned dark green and then black. The broth beneath it turned a golden yellow. This color, Fleming realized, was the result of the mold killing the bacteria in the broth. He tested the broth on the staphylococci. It proved as effective in killing the **germs** as the mold itself. Fleming experimented with weaker and weaker solutions of the broth. The staphylococci were wiped out even when the mixture was 500 parts water to one part broth. (Later studies showed that one part in a million would work as well.)

He tested the broth from the mold on different kinds of bacteria, including streptococci. Lysozyme had only acted on the least dangerous microbes. The broth from the mold, Fleming found, was able to interfere with the growth of bacteria responsible for causing the most serious diseases.

Fleming identified the mold as a member of the *Penicillium* family and, therefore, named his germ-killing substance penicillin. Then he dealt with the question of whether other molds would be as effective in dealing with germs as penicillin seemed to be. The **bacteriologist** began looking for different varieties of molds. This made Fleming the butt of much good-natured humor at the hospital. He would ask people for their moldy old shoes and collect mold from bread crusts and from the dank underside of porches. He tested these molds on the same germs that penicillin had wiped out.

None of the other molds had the same effect.

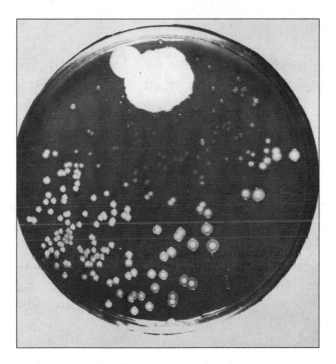

Alexander Fleming's own photograph of a penicillin mold. Penicillin was the first of many antibiotics—drugs that help doctors deal effectively with infectious diseases.

Penicillin was unique. Now Fleming set about making large quantities of the yellow broth for research purposes. He prepared hundreds of culture dishes, testing to see on which day of growth the penicillin was most effective, at which temperature, and in which particular solution of broth.

When these questions and others were resolved, Fleming had determined the most effective strain of penicillin in destroying microbes. Now he tested his "mold juice" on live animals. He injected rabbits and mice with the broth. They were unaffected by the same doses that killed staphylococci. In other words, the penicillin killed the germs without affecting the blood cells and living tissue of the animals.

This was the point at which he successfully treated Stuart Craddock's infected sinuses. Soon after, he used penicillin to treat the infection of a woman whose leg had been amputated. The infection was so far advanced, however, that the penicillin was not effective. There was a reason for this, as Fleming's later experiments with animals showed. It took penicillin up to several hours to destroy the microbes it affected. But when it came into contact with blood, it lost its effectiveness in about 30 minutes. Bacteria, such as staphylococci, worked their way into living flesh to infect the body. Penicillin lost its strength before it could penetrate the tissues. It would work as a local antiseptic with surface conditions like open sores or sinus infections, but it would not reach the germs it might kill when they were beneath the surface.

At this point, Fleming's penicillin lacked stability because he could not extract it in its pure form from the solution in which it grew. "Extraction and concentration," according to Craddock, who worked with Fleming on penicillin, "[were] essential" before it could be used to fight disease. But Fleming was a bacteriologist, and the procedure that Craddock described required a really good research chemist.

TELLING THE WORLD

Fleming was sure that extraction could and would be done. He was convinced that penicillin would become an important new tool in fighting disease germs. On February 13, 1929, he presented his findings to London's Medical Research Club. When he finished, there was a dreadful silence. Not one member of the audience commented on his findings or raised any questions about them. To Fleming it was a "frightful moment." A major discovery was being ignored.

He was not, however, the sort of man to give up. In June he published an article on penicillin in the *British Journal of Experimental Pathology*. At the end of the article, Fleming suggested that his discovery "may be an efficient antiseptic for application to, or injection into, areas infected with penicillin-sensitive microbes." Like his speech to the Medical Research Club, it received little attention. Today, the article is considered a classic in the field of medicine. It would be 13 years, from 1928 until 1941, before Fleming's discovery would become a major tool in the fight against harmful bacteria.

Before publication of the article, Sir Almroth Wright, the head of the Inoculation Department at St. Mary's Hospital, had objected to Fleming's claims. He was Fleming's boss and had the authority to approve or disapprove any paper that Fleming published. Committed to **serum,** a fluid containing a weak strain of bacteria injected into a person to create an **immunity,** as the means of fighting diseases, Wright thought Fleming's claims about his mold were exaggerated. But Fleming stood his ground. Wright reluctantly gave in, and the article was published as written.

One of the editors who accepted Fleming's paper for publication was an Australian-born pathologist named Howard Florey. Together with Ernst B. Chain, a young Jewish chemist who had fled Nazi Germany, Florey would make Fleming's dreams for penicillin a reality. The three men would make penicillin "the first of a series of antibiotics ... that have revolutionized medical practice." By 1946, penicillin would be widely available, and such natural and synthesized penicillins as ampicillin, carbencillin, and pipercillin would be developed later. All of these medications would help cure many serious diseases.

Alexander Fleming discovered the penicillin mold in 1928. Large-scale production of the penicillin drug did not occur until World War II.

CHAPTER TWO

THE EARLY YEARS

Alexander Fleming was born on August 6, 1881, at Lochfield Farm in the lowlands of Scotland, where the counties of Ayr, Lanark, and Renfrew meet. It was a region where sheep grazed over rolling hills ringed by distant mountains lost in the haze. The mists were thick on the moors, and the sun had to search to find the heather or the treasure of a bluebell. Even today, there aren't many people in the region.

Hugh Fleming, Alexander's father, rented the 800-acre Lochfield Farm from the Earl of Loudon. The 60-year-old widower already had four children—Jane, Hugh, Tom, and Mary— when he married Alexander's mother, Grace Morton. Together, the couple had four more children: Grace, John, Alexander, and Robert.

Grace was a loving mother to her children and her stepchildren. She was helped by her oldest stepdaughter, Jane, who was in her twenties when Alexander was a toddler. Jane called him Alec, a nickname that stuck, and he adored her. Shortly after Alec's youngest brother, Robert, was born, Jane came down with smallpox. Grace nursed Jane day and night, but at that time no **vaccine** for the disease had yet been discovered. Jane died.

Another tragedy followed. In 1887, when Alec was six years old, his father had a stroke. He could not move his legs and lingered for a year, confined to his bed or a chair by the fire before he finally passed away. Alec's oldest stepbrother, Hugh, took over the management of the farm. The family's devotion to one another and Grace's strength and love helped them deal with their father's death. The closeness between Alec and his brothers and sister and step-siblings would last throughout his life.

Alec and his brother Robert were particularly good friends. Robert was two years younger than Alec, and he followed Alec's lead in doing all the things that Scottish farm boys did a century ago, including trout fishing by guddling (the Scotch word for "tickling"), which involves leading the trout with a hooked worm and then landing the fish with a lightning-quick grab of the hand.

With the help of an old collie, the boys also caught rabbits with their bare hands. Their mother would cook a delicious rabbit stew with cabbage and other garden vegetables. The hides would be saved to make fur socks and mittens against the brutally cold North Atlantic snows that covered the lowlands in winter.

In the spring, Alec and Robert collected wild bird's eggs and sold them to a local grocer for pocket money. Together they explored the wild moors and hills and valleys of Lochfield. His boyhood taught Alec a love of the land he would have always. "As boys we had many advantages over the boys living in the towns," he recalled in later years.

"As boys we had many advantages over the boys living in the towns."

Along with other children from nearby farms, Alec and Robert played a dangerous game of downhill rolling. The hills of the lowlands are steep and the cliffs break off sharply. At the bottom there may be swift-moving streams or whirlpools. The game was to pick one of these downhill slopes and see who could roll down it fastest and farthest. Alec was a fierce competitor (as he would be all his life) in this game.

One day, on a dare, he rolled down a particularly steep hill that ended abruptly with a sheer drop into a canyon of rough-cut boulders. The other children gasped and cried out as he picked up speed, rolling and bouncing faster and faster. Seemingly, Alec had completely lost control of his plummeting body. Somehow, however, at the last minute, he turned his body to brake his speed and stopped just short of the edge of the cliff. As the other children ran down to him, he picked himself up, dusted himself off, and said, "Ah cam doun tha quick."

It was typical of both an accent and an attitude that would carry over into Alec's adult life. His speech would never lose its Scotch burr, and during his years in England he would make no effort to speak as the English spoke, even when it got in the way of his becoming a successful lecturer. In the Scotch manner, he would always use as few words as possible and make them count. Whatever his accomplishments—whether rolling down a hill or discovering penicillin—he would be modest about them, and always there would be a glint of humor in his eyes.

SCHOOL DAYS

One of Alec's teachers, Marion Stirling, remembered him "as a dear little boy with dreamy blue eyes." She taught in the one-room school known as Loudon Moor, which Alec started to attend when he was five years old. In good weather Alec walked barefoot to the school because feet could be dried, while muddy boots and stockings remained wet all day. Of course he didn't go barefoot in winter, and on those icy days his mother gave him baked potatoes hot from the oven to put in his pockets to keep his hands warm. Later he would eat the potatoes for lunch.

Alec attended the Loudon Moor school for five years. When he was ten years old he transferred to a more advanced school in the town of Darvel. He was an excellent student but not always a punctual one. One day in the school yard, racing around the corner of the building to make a class, Alec crashed into a boy heading the other way. Alec's nose was broken, and it would remain flattened out for the rest of his life.

When he was twelve years old and learning at a much faster rate than most of his classmates, Alec transferred to Kilmarnock Academy. Founded in 1633, the Academy had a proud Scotch tradition. The bustling town of Kilmarnock, with its population of 30,000, was a new world to the farm boy, but the world that unfolded to him at school was the one to capture his attention and shape his life.

The headmaster was ahead of his time in encouraging science studies in his school. Students were required to take two science subjects each year. The emphasis was on such theoretical topics as inorganic chemistry, physical geography, magnetism and electricity, light and sound, and physiology.

Alec profited in two ways from these studies.

He gained a wide knowledge of the sciences and an understanding of how they were interrelated. He had an excellent memory, and years later he could still quote from some of the Kilmarnock textbooks and lectures. Perhaps more importantly, he developed habits of imagination and logic that led him to raise questions and to persist in seeking answers.

His time at Kilmarnock was, in a way, the beginning of Alexander Fleming as a scientist. The road was opened to him. There might be detours along the way, but his life's journey had begun. His first steps led to London.

OFF TO LONDON

London, the sprawling capital city of England, was a far different place at the turn of the century when young Alec Fleming arrived there than it is today. It was part of a different world—a world without computers or CD players, television or radio, airplanes or automobiles. The streets were made of cobblestones, the unpaved sidewalks crowded with pushcarts selling meat and vegetables. Horse-drawn wagons crowded the avenues, and when the traffic thinned, coaches would hurtle past at such great speeds that people on foot would be forced to jump out of their way. The air was thick with the smell of horses and manure.

Pedestrians, wagons, and carriages crowd London Bridge in 1892. Fleming arrived in London three years later to study at the Polytechnic Institute. His family wanted him to pursue a career in business.

The city's first underground subway had just been built, and the steam from the coal-powered trains billowed up from the grates positioned along the streets. The first London house that Alec lived in was situated over the subway line. Every time a train passed, the house would shake and the ominous rumble would usher in gusts of smoke through the floorboards.

This house on Marleybone Road was the home of Alec's stepbrother Tom, who had become an oculist, or eye doctor, and had set up practice in London. Tom's sister, Mary, kept house for him, and Alec's older brother John, who had found work in London, also lived there. Soon after Alec arrived, his younger brother, Robert, joined him.

Both boys were overwhelmed by London. If Kilmarnock had introduced them to city life, London dazzled them with the seeming chaos of a bustling metropolis. They would never again feel quite so Scotch as during those early days with the buzz of British accents, ranging from the working-class Cockney to the sniffs and nasal drawls of the gentry, drowning out their lowland burrs.

By coming to London, Alec and Robert were maintaining a family tradition that was strongly Scotch. One word sums it up: education. Scotland was then a poor country of farmers; England was a rich country where commerce and new industry flourished. The opportunities for both work and higher education were much greater in England than in Scotland. In the Fleming tradition, education was the requirement for a successful life. Alec, 14, and Robert, 12, enrolled in the distinguished Polytechnic Institute. The name is misleading: Polytechnic Institute was really a commercial school. The older Flemings wanted to prepare the boys for careers in the business world.

Despite the interest in science that Kilmarnock Academy had sparked in Alec, he bowed to the influence of his beloved family and went along with preparing for a business career. During his first two weeks at Polytechnic, the school determined that Alec was two years ahead of his classmates and moved him up four classes.

Now the youngest in his class, Alec's youth, combined with his Scotch accent and country-boy manners, made him the butt of jokes by his schoolmates. He became somewhat withdrawn, but his dry Scotch wit—expressed in a few murmured words—earned him the liking and respect of many who teased him. He did not shine socially as he did in his schoolwork, but he was not unpopular.

High marks came easily to Alec. He had an excellent memory, but his greatest asset was his active imagination. "He was always thinking something up," according to Robert. "But if it didn't work, he lost no time in modifying it or discarding it altogether and thinking up something else." Some day that "something else" would save millions of lives. Alec's imagination was encouraged by his stepbrother Tom, who often set up competitions for his brothers in such subjects as geography, history, mathematics, and science. Each of them put up a penny, and the one with the most correct answers to a series of questions won the pot.

Tom also urged the boys to exercise their bodies as well as their minds. He brought home a pair of boxing gloves. Each boy got one—the right-hand glove to the younger boxer, the left-hand glove to the older. The boys fought one-handed, using only the fist with the glove. The matches would start out as sparring but would almost always lead to slugfests. Tempers would be lost, and fists would fly. This so upset Mary Fleming that she finally put a stop to the brawling by throwing out the boxing gloves. After that, the brothers got their indoor exercise playing table tennis.

Alec's life wasn't all play though. His excellent marks in school had put him in line for a job with a major shipping company. At age 16 he left the Polytechnic Institute and became a junior clerk with the America Line, which operated some of the most luxurious passenger liners crisscrossing the Atlantic. He entered figures in account books, made copies of documents, and kept records of passengers' bookings.

It was a dull job, and he hated it. Nevertheless, Alec did what he would always do in his later career as a scientist. He dug in his heels and stuck with the job for the next four years. The only protest he made was mild, but it got him noticed. He began wearing a bow tie at a time when most men wore foulards, broad neckties that covered the expanse of shirt below the collar. Alec wore bow ties all his life, and they became his trademark among the scientists with whom he mixed.

While Alec was still in his teens and doing work he loathed, the Boer War broke out in South

Africa between Great Britain and South Africans of Dutch descent (the Boers). Early in 1900, Alec and his brother John enlisted in the London Scottish Rifle Volunteers. The following year their brother Robert joined them.

None of the Fleming brothers, however, saw action. Indeed, aside from some training exercises in the pouring rain, they rarely left London. Nevertheless, both Alec and Robert became active in their regiment. They were both crack shots and represented the Scottish Volunteers in marksmanship competition. They were both on the team that won the Daily Telegraph Cup, and Robert went on to represent Scotland in international competition. The Fleming brothers also joined the regimental water polo team, and Alec competed against St. Mary's Hospital Medical School, an institution that would play an important role in his future.

Alec loved the sporting life of the regiment. "The men of H Company," he would remember with a twinkle in his eye, "were self-opinionated, egocentric and obeyed no rules but their own." This was just the opposite of the restrained atmosphere of his job.

Shortly after his twentieth birthday in 1901, Alec was rescued from his dreary job when a well-off bachelor uncle died and left his estate to his relatives. Alec's share was 250 British pounds: the rough equivalent of $1,250, which was a very large amount in those days. Encouraged by his brother Tom, Alec decided to quit his job and use the money to go to medical school.

There were obstacles standing in his way. At age 20, Alec was two or three years older than most applicants and did not have the educational credits most schools demanded. He had never studied Latin, which was a strict requirement for admission to medical school. Determined, Alec hired a tutor to teach him Latin. He enrolled in the London College of Preceptors to make up his other credits. Then he had to pass a stiff examination covering a variety of subjects to be eligible for acceptance to a medical school. The prize given to the student who scored highest in the examination did not go to Alec; it went to the 15-year-old student whose overall score he had tied.

"Competition was the breath of life to him."

As in hill-rolling, boxing, table tennis, rifle-shooting, and water polo, Alec had once again shown himself to be a fierce competitor. According to a friend who knew him well at that time, "competition was the breath of life to him." Nevertheless, as medicine became his life, competing with people became far less important to Alec than the need to relieve their suffering.

CHAPTER THREE

BECOMING A DOCTOR

Along with 79 other would-be doctors, Alexander Fleming entered St. Mary's Hospital Medical School in 1901. He was 20 years old, a fair-haired young man with a permanently squashed nose and a penetrating blue-eyed stare that drew some people to him but which others thought insolent. He was shorter than average and self-conscious about it. "Tall people can do anything, go anywhere," he observed with humor, but also with envy. He was physically strong, well enough liked by his fellow students, but slow to form close friendships.

The world of medicine Alec was entering was one of constant change and some turmoil. More new ideas had swept over it in the past 50 years than in the preceding 20 centuries. Among the first was a simple enough rule proposed by a Hungarian physician named Ignaz Philipp Sem-

Fleming (front row, right) listens to a lecture at St. Mary's Hospital Medical School.

melweiss in a book he wrote in 1861, some 20 years before Fleming was born. Semmelweiss's revolutionary suggestion was that doctors should wash their hands before examining patients and that surgeons should do so before operating.

GERM THEORY

Little was known of germs when Semmelweiss suggested this. He had been practicing in a hospital in Vienna when he noticed that the death rate among women in maternity wards was 25 percent higher than among poorer women who gave birth at home. The former had their babies delivered by doctors, the latter by midwives—women without medical degrees. These deaths were due to puerperal fever, commonly known as childbed fever.

Semmelweiss realized that the midwives, who like most poor women of that time did much cooking and cleaning, washed their hands frequently to rid them of the odors of their labors. Doctors, on the other hand, often went to the maternity wards straight from treating diseased patients or dissecting corpses. They were too busy to wash their hands between activities. The Viennese medical community reacted indignantly to Semmelweiss's suggestion. Look! Their hands weren't dirty! It was an insult to ask them to wash their hands. They ran him out of Vienna. In the London hospitals of 1901, doctors were still not required to wash their hands. (It wasn't until the 1930s that it became a requirement in New York City hospitals.)

Semmelweiss had been followed by the French scientist Louis Pasteur, who proved that some germs cause certain diseases and who developed the first ways to fight them. He showed that by identifying the specific germ that caused a disease and breeding a weakened species of that germ, it was possible to make a vaccine that could be injected into animals and make them immune to the disease the germ caused.

Meanwhile, German bacteriologist Robert Koch had been studying the 2,500 kinds of bacteria that live in the human body. He identified the types of bacteria that caused tuberculosis and cholera. He showed that bacteria carried by fleas caused bubonic plague.

In Edinburgh, the high rate of deaths in surgery persuaded surgeon Robert Lister that germs existed everywhere—in the air we breathed as well as in the flesh of patients. He developed a carbolic acid solution to kill the germs on flesh during surgery without destroying the flesh itself. He introduced a carbolic acid spray that filled operating rooms with a fine mist meant to kill the germs in the air. Although the procedure was very unpleasant and not completely successful, the rate of deaths during surgery was dramatically reduced. Nevertheless, Lister's search for an antiseptic that would kill all germs without harming the patient did not succeed.

The combined work of these scientists and others came to be known as germ theory. In

*Robert Koch (1843–1910) was one of the
19th-century scientists who demonstrated
that microorganisms cause many diseases.*

London at the turn of the century it was still by no
means accepted by everyone in the medical pro-
fession. One physician who did accept it and
enlarged upon it was Sir Almroth Wright. In 1898
Wright developed a vaccine to prevent typhoid.
Wright became Professor of Pathology at St.

Mary's the year after Fleming entered medical school. He had just lost a fight to have his vaccine used in the British army. Many doctors at the time opposed the idea of injecting germs in people to make them immune to the germs that made them sick.

Unlike Pasteur's vaccines, the germs Wright used were already dead. Alec Fleming, who would become one of Wright's steadfast followers, explained the importance of this. He wrote that living germs might work in vaccinating animals, but were not, "owing to certain risks, suitable for the treatment of humans." Alec considered Wright's development of a dead-germ vaccine that eliminated the danger a major step forward in fighting disease.

By the time Wright came to St. Mary's, Fleming had already established a reputation as a star student. During his first year he won a scholarship prize, which paid all his tuition fees The following year he won prizes in Anatomy and Physiology. When he spoke of this, Alec credited the time during which he had dropped out of school. He thought it was a great advantage to have taken time off from books to experience "the school of life." That school-of-life seasoning may have helped Alec to be appointed a prosecutor, the student who helps to prepare a dead body for dissection and who demonstrates with a pointer and charts and pictures during a professor's anatomy lectures. The following year, 1904, he won the Senior Anatomy Prize, the Histology Prize (for his work with a microscope), and the Junior General Proficiency Prize. He was now taking classes with Professor Wright, but he had decided against becoming a bacteriologist. Alec was extremely skillful with his hands, and in 1905 he took and passed the examination to enter the Royal College of Surgeons.

At St. Mary's he had received hospital training in the wards, which included performing the most menial tasks from emptying bedpans to changing bandages. In the emergency room he treated minor ailments, dressed wounds, and even pulled teeth. He came into contact with many childhood diseases—measles, mumps, scarlet fever—for which there was no real treatment yet. Vaccines for these diseases were still in the process of being developed. Streptococcus infections were a major killer at the time. Alec saw firsthand the suffering that resistance to the acceptance of germ theory was causing.

In 1906 Alec passed his final examinations at the Royal College of Surgeons. He was now a qualified doctor. He did not have the money, however, to open an office and set up a practice. With all his accomplishments, the plain fact was that he needed a job. His ambition was to be a surgeon, but the only offer that came his way was a far cry from that. He was offered a position as junior assistant in Sir Almroth Wright's Inoculation Department at St. Mary's.

Alec accepted the position with the understanding that it was only on a temporary basis. He was determined to be a surgeon. He looked at the job as just another detour in his career.

THE MAGIC BULLET

When Alexander Fleming joined Sir Almroth Wright's staff, the focus of their work was phagocytes. A phagocyte is a particular type of **white blood cell** that fights against infection-causing germs. White blood cells are the most important part of the human **immune system**—the body's own defenses against disease.

If you cut yourself and the wound becomes infected, phagocytes come to the rescue. They multiply and "eat" the microbes. (Microbes, also known as bacteria, are germs.) Wright had discovered that in human blood there is a liquid, now know as *serum,* that makes the phagocytes eat the bacteria more greedily.

Microbes can be placed on a slide and studied under a microscope. Wright coated the germs in a drop of blood with serum, fed them to the phagocytes, and observed the results. The phagocytes

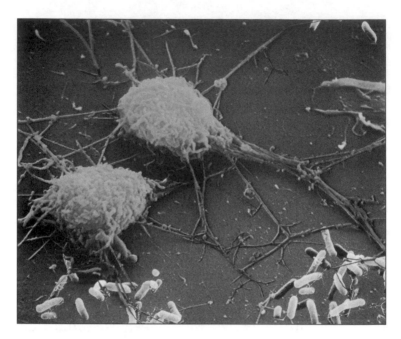

*This electron micrograph shows a macrophage
(a type of phagocyte) attacking an E. coli bacteria.
A phagocyte is a kind of white blood cell that
destroys bacteria.*

ate more of the serum-covered germs than they
did germs not coated with serum. In other words,
the serum increased the appetite of the phagocytes for the microbes.

Wright could actually count the number of
microbes eaten by the phagocytes. He compared
them with the usual number of microbes in a cell
before infection takes place. By comparing the two

microbe counts, Wright could prepare a chart that would indicate how much serum would be needed to stimulate the phagocytes to eat the germs in a specific kind of infection.

Alexander Fleming's first chore in the Inoculation Department at St. Mary's was to help gather this information. He put in as much as 16 hours a day preparing slides, peering through a microscope, counting microbes, and recording the results. It was both boring and exacting work. His skill and devotion impressed the demanding Wright.

Recognized in his field as a leading research scientist, Wright had a reputation as a hard taskmaster and a stubborn man of strong, sometimes harsh, opinions. Fleming didn't share Wright's views on many topics, but he did admire his boss's scientific genius. He profited by Wright's conviction that "before making a decision on any subject, a man should always have numerous conversations with those who are expert in it." On the other hand, Wright could be quick to jump to a scientific conclusion and would stick to it stubbornly despite evidence to the contrary. Playing on his full name, Sir Almroth Wright, his critics called him "Sir Almost Right."

A colleague remembers that Fleming dealt with Wright's stubbornness quietly and sensibly, the only sign of strain being that his Scotch burr thickened. He remembered that when Wright got carried away, "Fleming was always courageous enough to say, quite calmly: 'That won't work, sir.' Wright would repeat his argument even more forcibly. Fleming would listen, without interrupting him and then, quite simply, say again: 'That won't work, sir.' And it didn't."

"That won't work, sir."

On the whole, however, Fleming agreed with Wright's serum theories. In 1907 and 1908, Alec published two papers on them in medical journals, one written with Wright, the other with another researcher. During this period, however, he was also practicing medicine apart from Wright.

In May 1908, he qualified as a surgeon. He became Casualty House Surgeon at St. Mary's while continuing to work with Wright in the Inoculation Department. He performed minor operations and assisted at major ones.

Busy as he was, however, Alec always made time for recreation. Bashful with women, this usually involved other men. He took up snooker (a variety of pool) and billiards and became expert at both. He and his brother Robert played golf regularly. Alec also became an accomplished glass blower and an amateur painter. He joined the Chelsea Arts Club, where he was "elected 'Honorary Bacteriologist,' on the condition that members would get free treatment from him."

Despite this kidding, Fleming was never anything but deadly serious about his work. This was true when in 1908 he began focusing his attention on acne, a condition that affected millions of young people but that other medical researchers at the time considered of little importance. Skin glands produce an oil that lubricates skin cells. When too much of the oil is produced, the skin erupts. Sores and blisters appear. This is acne. It's caused by different types of bacteria that act on the skin glands to overproduce the oil.

Fleming studied the pus from acne sores under a microscope. He identified three kinds of germs that caused three varieties of acne. By taking blood from an acne sufferer and examining it, he could tell which kind of acne the patient had. He then began to breed weak strains of acne bacteria in order to make vaccines to treat the condition.

In April 1909, Alec published the results of his research in *Lancet*, a British medical journal. "We may take it as definitely proved," he wrote, "that in localized infection, when one inoculates the patient with appropriate doses of a carefully prepared vaccine derived from the infecting organisms, one obtains a beneficial effect." The vaccine didn't cure the disease. It did, however, make the sores go away. In some cases they returned; in some cases they did not. It wasn't a complete remedy, but it was progress.

Fleming next turned his attention to a far more serious disease: **syphilis.** Passed from person to person by sexual contact or from an infected mother to her baby at birth, syphilis was considered to be almost always incurable and fatal in the early 1900s. In 1909 this changed with the work of German chemist Paul Ehrlich. He was a friend of Sir Almroth Wright's, but their approach to disease was very different. Wright believed in isolating

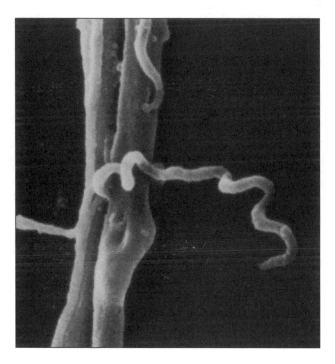

This electron micrograph shows several treponemes, or syphilis bacteria, attached to testicular cell membranes. Syphilis was almost always incurable in the early 1900s.

germs and creating vaccines, and he had passed this conviction on to Fleming. Ehrlich, on the other hand, thought that if a germ could be identified, a chemical could be developed that would kill it.

The microbe that causes syphilis is called *Treponema pallidum.* Between 1905 and 1907, various arsenic-based drugs had been developed to kill the syphilis germs. They were called **atoxyls.** The name literally means that the substances aren't poisonous, but, in fact, they all were. Ehrlich set himself the task of determining the minimum dose that would destroy the microbes and the maximum dose that could be given without killing the patient.

He put to death thousands of mice and guinea pigs in these experiments. Altering the strength of the atoxyls, Ehrlich tried one combination after another. These were numbered. Finally, in May 1909, the tests conducted with version number 606 got results. The germs were destroyed without killing the syphilitic animals. When formula number 606 was applied to the open sores of rabbits with syphilis, complete cures were achieved in three weeks. The medical establishment hailed formula 606 as Dr. Ehrlich's "magic bullet," a substance that attacks specific bacteria in the body without harming anything else in the body. Ehrlich called it **salvarsan.**

Wright had doubts. He was committed to the

Paul Ehrlich (1854–1915) developed the arsenic drug salvarsan, which cured syphilis.

germ theory and mistrusted treatment of disease with drugs. What might work on mice and rabbits might act entirely differently on human beings. Despite his misgivings, when Ehrlich came to lecture in London, Wright introduced him to

Fleming. Ehrlich had brought some of the salvarsan with him, and he gave some to Fleming to use on patients suffering from syphilis. Immediately, Fleming realized the drawbacks to using 606 on people.

When the drug was exposed to the air, it was oxidized—a process that altered its chemical makeup. Also the 606 had to be injected directly into a vein. Intravenous therapy was a very new process to most doctors. The equipment was crude, and great dexterity was needed to give the injections, which could be long-lasting and painful. Fleming was one of the few doctors who had mastered this procedure.

Dr. G. W. B. James recalled watching Fleming treat a patient with salvarsan. He described him "setting up a glass reservoir containing a yellow fluid, inserting a needle into a vein on the patient's arm, and running the fluid directly into the bloodstream." James was struck by "the rapidity with which the 606 took effect."

MAKING A NAME

News of Fleming's successes in treating syphilis spread. Other doctors sent him patients suffering from the disease. He found himself with a growing private practice and a reputation as the syphilis doctor. A famous cartoonist, Ronald Gray, drew a caricature of Alec in the uniform of the Scottish Volunteers. The figure carried a giant hypodermic syringe in place of a rifle. The caption read, "Private 606."

This celebrity status gave Alec a sort of back-door entry into the society of notable Londoners. He began to go out more. He learned to dance and even attended a fancy dress ball costumed as a little girl. He became relaxed in mixed company. Although he was never good at small talk—Alec was often shy and could be brusque—he did have a quiet wit that drew people to him. Many who got to know him, women in particular, found him to be a warm and sensitive person.

In those years just before World War I, Alec was in his early thirties. He continued to live in the London house that he shared with his family. An eligible bachelor, much in demand as an extra man at dinner parties, he showed no sign of becoming involved with any of the fashionable young women who encouraged him. But the war that began in 1914 and lasted four long years would change that, as it would change so many things.

CHAPTER FOUR

DOCTOR AT WAR

Ten million soldiers were killed in World War I, the conflict in which England, France, Italy, and Russia—and later the United States—fought against Germany, Austria-Hungary, Turkey, and Bulgaria. Twenty-one million were wounded. A million or more were missing and presumed dead. Of the 10 million known dead, most did not die on the battlefield but rather in the field hospitals where they were taken for treatment. They died of wounds that had become infected. Much of the infection was due to the various poison gases that both sides used in the war. The gas soaked into the uniforms of the soldiers. **Tetanus** poisoning, caused by a germ that enters the body through wounds, was the major cause of death.

An estimated 20 million people, twice as many as those killed by the war, died in the Spanish influenza **epidemic** that swept the world as the war was ending. Like the germs of war, the germs of the epidemic came under the microscope of Alexander Fleming. He began to consider the possibility of a substance that might rob all germs of their ability to cause infection, disease, and death.

But that came later. At the beginning of the war, Alec was kept much too busy to think in such large terms. Sir Almroth Wright had put his department at St. Mary's to work producing the typhoid vaccine he had developed. Years earlier, the British War Department had rejected the vaccine, but now some 10 million doses were prepared for the soldiers going to war. Fleming was Wright's chief assistant in carrying out the program.

The typhoid vaccination program was the first recognition of the dread diseases that might be spread by war. Later there would be a measles epidemic as well as the plague of Spanish influenza. The danger was recognized early by Sir Alfred Keogh, who was in charge of the British Army Medical Services. "In this war," he realized, "we have found ourselves back among the infections of the Middle Ages."

A FIELD LABORATORY

Keogh assigned Wright to set up and run a medical facility "to study the bacteriology of wound infection and the best methods for overcoming it."

The unit would be a field hospital in Boulogne, a seaport in northern France. Wright was made a colonel in the Army Medical Services, and Fleming, who went with him to France, was given the rank of lieutenant.

The hospital was housed in the Casino, a once swanky establishment that was now in bad shape. The plumbing leaked, and the ventilation was bad. From the first they were short of supplies and medical equipment. Fleming somehow put together his own stoves and pumps to supply water and laboratory devices. His first work in the hospital was that of a talented handyman with a knack for tinkering. Nevertheless, he would later call the Casino "one of the best laboratories" he had known.

Wards were set up with rows of cots for the wounded. From the first the staff was overwhelmed by their sheer numbers. The casualties of battle in World War I, due to the technical advances in weapons, were greater than anyone had anticipated. Soldiers might lie wounded for days before receiving patchwork battlefield treatment by medics. Shortage of ambulances and drivers caused further delays in their being transported to hospitals. When they arrived, there were still more delays while they waited their turn to see an overworked doctor or to be taken into an operating room, where a bleary-eyed surgeon might try to patch them up.

*Fleming (far right) poses with
fellow officers of the Royal Army
Medical Corps during WWI.*

The initial treatment on the battlefield or in
the hospital was always the same. Once the bleed-
ing had been stopped, the wounds were cleaned,
swabbed out, and bathed in antiseptics—the
germ-killing substances developed by Robert
Lister. Ever since Lister, the accepted theory had
been that these antiseptics killed the germs and
kept the infection from spreading.

At first Fleming did not challenge this. His initial task was to identify the germs that caused the infections. He took swabs from wounds before surgery, during surgery, and after it was over. Sadly, the last swabs too often came from dead bodies. He also examined bullets, shell fragments, and clothing.

Based on this investigation, Fleming deter-

Wounded soldiers await treatment at a British field dressing station during World War I. During the war, Fleming labored under adverse conditions to develop procedures to save the lives of the wounded.

mined that the main cause of infection was not the wound itself but the materials that came in contact with it. Many wartime wounds resulted in tetanus, and the germs that caused it thrived in porous material—chiefly cotton and wool, the fabrics used in the soldiers' clothing. This was particularly true when poison gas had seeped into the fabric. He concluded that the uniforms of the wounded men were the major cause of spreading the infection.

TREATING WOUNDS

In 1915, Fleming wrote two papers on his findings for the medical journal *Lancet*. He "described the bacteria responsible, their frequency, origin and conditions of growth." Then he discussed the treatment of the wounds. This would spark a major debate among battlefield surgeons.

Alec reported that while he had been trained to use antiseptics to dress wounds, his experience in war was that they were not effective. He had seen that peroxide, carbolic acid, boric acid, and other antiseptics did not kill all the microbes. The theory had been that some microbes were killed, which was better than not using antiseptics at all. Now he had discovered evidence to the contrary.

He had found that in swabs from fresh wounds that had not yet been treated, the phagocytes were actively killing the tetanus germs. But in swabs taken from wounds treated with antiseptics, the phagocytes were dead or dying, and the germs were flourishing. The interaction was complicated by the brutal nature of the wounds and the filth of war, but the conclusion was obvious: The antiseptics were killing more phagocytes than germs.

Fleming took his findings to Wright, who acted quickly but impulsively. He began a campaign to have army surgeons stop using antiseptics in the treatment of wounds and to use a strong, sterile salt solution that would promote the growth of phagocytes instead. This worked well with most wounds, but not so well with those that were deep and jagged or heavily infected. The army doctors split into two camps—for and against the use of antiseptics.

Wright sent a memorandum to the War Office condemning the use of antiseptics for battlefield injuries. Medical officers who outranked him opposed this, as did the president of the Royal College of Surgeons. There was a campaign to have Wright sent back to England. Wright appealed to British Prime Minister Lloyd George to back him up. In the end, Wright was not removed, but the practice of treating wounds with antiseptics continued all through the war.

Fleming had come under attack along with

Sir Almoth Wright works in his laboratory.
Fleming often disagreed with Wright, but
respected him as a scientist.

Wright. He would always be coupled with Wright in the eyes of the scientific community, and much of the criticism of Wright for his insensitivity and arrogance would rub off on him. This was unfair, for the two men were nothing alike. They did, however, share a view based on their wartime experiences: The way to overcome infection was not to apply antiseptics but rather to defeat germs by stimulating the growth of phagocytes in the patient's own body. The difference between them was that Fleming, unlike Wright, also saw the possibilities of chemical agents, such as Ehrlich's magic bullet, as major weapons in the fight against germs. This open-minded approach would lead Fleming to his greatest discovery.

CHAPTER FIVE

LOVE AND MARRIAGE

In 1915, after one truly terrible year of war, Alec returned to London on leave. His family introduced him to a new neighbor by the name of Sarah Marion McElroy. Everyone, including Alec when he got to know her, called her Sareen. Born in County Mayo, Ireland, Sareen was lively and as enthusiastically Irish as Alec was gently and patiently Scotch. From the first it was an attraction of opposites.

There was much to attract Alec. Sareen's vitality was heightened by tumbling blonde hair, a fair complexion, flushed cheeks, and dancing eyes. Her beauty dazzled him, and he had little in the way of looks to offer in return. It didn't matter. Along with her beauty,

> **"Alec is a great man, but nobody knows it."**
> **—Sareen Fleming**

Sarah had great insight. She saw past his squashed nose and sometimes stern face to what he was and what he could be. "Alec is a great man," she told friends, "but nobody knows it."

Sareen was not, however, a woman who would ever be content to sit at his feet in awe. Like her twin sister Elizabeth, she had been trained as a nurse. Before coming to London she had worked for a famous surgeon, Sir Thornby Stoker, in Dublin. Through him Sareen had met many famous people, including the writers George Moore and William Butler Yeats. She had come to London to open a private nursing home. It was a great success, and many famous artists and aristocrats were sheltered there. Sareen was very much her own woman.

Their love for each other blossomed quickly. The war, and the knowledge that Alec would have to return to it, may have pushed them faster into marriage than would have been the case in peacetime. Even so, they were slowed down by Alec's shyness and his Scotch reluctance to act hurriedly. Sareen understood his feelings even when he didn't put them into words. She provided the encouragement that nudged him into proposing. Once that was done, Alec relaxed and accepted the whirlwind plans necessary for the marriage to take place before he returned to the war. They were wed on December 23, 1915.

Sareen's twin sister, Elizabeth, was at the wedding. Elizabeth, who had been married to an Australian and was now a widow, had gone back to practicing as a nurse. Although she and Sareen looked alike, Elizabeth was withdrawn and quiet. Her remoteness did not discourage Alec's older brother John, though. Elizabeth and John soon fell in love, and they were married not long after Sareen and Alec's wedding. The brother-husbands and the sister-wives complemented each other. Alec, in particular, would find Sareen's outgoing and generous nature a blessing in the years to come. She understood him in ways that few other people did.

Sareen recognized that his work was solitary in nature and required him to be alone. She didn't resent this. Instead she adapted, making sure that she had her own things to do while he was occupied. To use a modern phrase, she gave him space but never yielded her own. She saw that his ability to focus on a problem was at the core of his greatness. She not only respected his work, but the day would come when she would sell the successful nursing home she had founded in order to get the money for Alec to continue his researches. Money was never uppermost with Sareen. Alec's career and her own independence always were.

Sareen would sacrifice without ever giving up her independent spirit. When money pressures forced them to get along without servants, Sareen would perform the household chores herself. She would do this without complaint, with dignity, but never with the pretense that she found them pleasant or rewarding. What she would find pleasant and rewarding, as would Alec, was gardening. He had a green thumb. According to a friend, he "just went up to some improbable tree when it was in full leaf, wrenched off a branch, stuck it in the ground, and lo and behold, it produced roots."

The time would come when Alec and Sareen would have a country home called The Dhoon. It would have two greenhouses and several gardens—all of which they planted and cared for themselves. Gardening was a touch point of their marriage, the thing that would bring them more closely together than anything else except the birth of their son.

They would enjoy going to antique shops together to find furniture for The Dhoon. Sareen would take up golf, and they would stroll the links hand in hand, with Alec offering occasional advice on her posture and swing. He might tease her gently. She might react more strongly. Large blue eyes twinkling, he would always back off.

They would practice putting on the lawn, and in the evening they would play croquet. When the sun went down, they would finish the game by candlelight. Always they would play well together, but always Alec would work alone.

But that was all in the future. Now, shortly

Fleming poses with his first wife, Sareen.

after their marriage, Alec had to part from Sareen for the first time. He had to leave his bride, cross the English Channel, and go back to France, where the darkness was lit not by candlelight but by the flares that preceded the cannon's roar. Alec had to return to the war.

CHAPTER SIX

BLOOD, BONES, AND INFLUENZA

The war dragged on through 1916. On the western front, where the English and French fought the Germans, the battle lines stretched across Europe. In the course of attacks and counterattacks, small advances and small retreats, hundreds of thousands of lives were lost. Little territory, however, was actually exchanged. When it was, it was frequently retaken by yet another bloody assault. The war was at a stalemate—a horrible, blood-soaked stalemate.

The military hospitals were overwhelmed. Fleming and the other doctors were hard put to keep up with the numbers of wounded who poured in every day. When the United States entered the war in April 1917 and began to shatter the stalemate, the fighting became more aggressive and violent. The casualties mounted, and the ability of the hospitals to deal with them became even more strained.

Fleming, Wright, and the other doctors at the Casino in Boulogne were badly shaken by "the terrible and tragic fate of so many of the wounded," wrote Oxford professor Gwyn Macfarlane. "In later years they could hardly bear to recall their desperate and often unavailing work in the wards." During this time Alec was more engaged in treating the wounded and trying to save lives than in continuing his research. That changed early in 1918 when he was assigned as a bacteriologist to a hospital at Wimeraux, which had been set up to deal exclusively with wounds involving bone fractures.

In some cases the break in the bone isn't complete and the two pieces of the bone aren't separated. The bone is splintered. In other cases, it is a clean break. Either way, a broken bone is not usually a very serious injury. Children fall out of trees and fracture their arms or collarbones all the time. Athletes break their ankles or their legs. People slip in bathtubs and break a hip. The bones are reset, splints are applied to hold them in place, sometimes a limb is encased in plaster so that it can't move while it is mending. In time, healing takes place, and the bone is as good as new. But there is a difference with wartime wounds involving bone fractures.

The filth of World War I turned simple fractures into potential fatalities. Such wounds became particularly vulnerable to **gangrene,** which was caused by a variety of germs called **anaerobes.** Oddly, anaerobes cannot grow where there is oxygen present. They exist in soil and in other material, but they are passive and usually have no effect on their environment.

Fleming observed that when anaerobes are "introduced into an animal body with ... a bullet" they become active, and "the toxins formed by

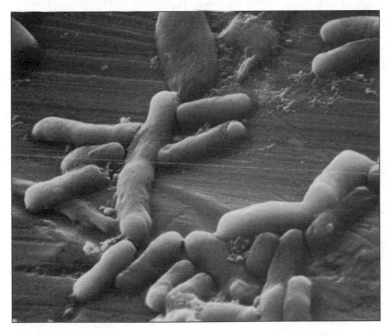

This electron micrograph shows Clostridium perfigens, *the bacteria that causes gangrene.*

them produce infections" such as "tetanus (lock-jaw), war-wound infections, gas gangrene." He noted that the anaerobes could be destroyed by providing them with oxygen and that if this was done quickly enough, the infections in bone fractures could be prevented.

Fleming devised a procedure for doing this. First it called for removal from the body of foreign objects, such as bullets, shell fragments, and dirt. Then it called for opening the wound fully before applying any antiseptic, or salt solution, so that the flesh affected by the poison of the anaerobes could be cleared away. This also exposed the anaerobes to the air. In this way their poisons were removed from the vicinity of the splintered bone, and they were themselves destroyed by the oxygen in the air. Only after this was done would the bone be set and the wound closed.

As a result of Fleming's recommendations, battlefield medical procedures were changed. The number of infections in war-wound fractures was reduced significantly. In the latter days of the war, many casualties of broken bones owed their lives to Fleming's new procedure.

PERFECTING BLOOD TRANSFUSIONS

Around this time he also began working on different techniques of blood transfusion. This practice was in its infancy at the time of World War I. The various blood groups had only been identified in 1901 by Austrian physician Karl Landsteiner. A reliable method for grouping those who gave blood with those who could receive it had only been in existence since 1908. Both of these advances would have been meaningless, however, had it not been for a means devised in 1891 that prevented blood from clotting.

Clotting would have meant that the blood did not stay liquid enough to be passed from one person to another. (There were no blood banks until World War II; blood had to go directly from person to person.) The means to prevent clotting involved adding sodium citrate, an anticlotting salt, to the blood. It was first used on the battlefield in World War I. The 1891 discovery that sodium citrate would prevent clotting had been made by Fleming's erratic and brilliant boss, Sir Almroth Wright.

Inexperience with blood transfusions was a serious problem for medics, nurses, doctors, and surgeons in World War I. They started out with little practice in finding veins. Because weather conditions, humidity, and temperature affected the anticlotting ability of the sodium citrate, medical workers often had trouble finding the proper concentration for the blood solution.

Fleming set about determining how the process of blood transfusions might be made easier and more uniform. He talked some soldiers into donating blood by getting them extra leave in return for volunteering. Then he experimented with the blood and with the sodium citrate. He ran tests to measure the flow of the blood under a variety of circumstances. At the same time, he worked on ways to make the veins easier to find and hold and to make the transfusions less painful for the patient receiving them.

The result of Fleming's experiments was the development of practical methods for blood transfusion. Along with A. B. Porteous, a physician who worked with him on this, Fleming published the results of their work in the *Lancet* after the war.

FIGHTING THE FLU

By the time Fleming returned to the Casino in Boulogne, the influenza epidemic of 1918 was raging. It had begun in Spain in May, some six months before the war ended. It spread with lightning speed throughout Europe and the United States. At first, although the number of people stricken was alarming, not very many died of the disease. By August, the epidemic seemed to be winding down. But then, in October, a second wave of the influenza infection broke out, and this time it was much more deadly. "Healthy young people could report sick with a mild fever on one day," wrote Gwyn Macfarlane, "and be dead on the next."

The cause of Spanish influenza was a **virus.** Different from other germs, a virus is about one-tenth the size of most bacteria. Its design is different and simpler—it has a core of **nucleic acid** and a coat of protein, sometimes with fat and carbohydrate material added. Whereas ordinary bacteria can live and reproduce independently, viruses are nonliving infectious particles that need living cells to reproduce. They "inject" their nucleic acids into host cells, which then reproduce the viruses many times. The living cells are killed when they burst, releasing multiple copies of the newly reproduced virus particles.

In 1918, the Spanish flu virus had not been identified. It was believed that Spanish influenza was caused by a virus known as *Haemophilus influenzae.* Only in 1919 did Japanese bacteriologists show that *Haemophilus influenzae,* although often present in the disease, did not cause it. A short while after that, a different virus was established as the cause.

Fleming was trying to deal with the day-to-day horror of the epidemic while at the same time trying to study *Haemophilus influenzae* in order to find out why fatal lung infections were so much more common now than in the earlier epidemic. He worked desperately to keep soldiers from choking to death, to relieve their breathing problems, to somehow stop the collection of fluid in their lungs—fluids that were drowning them. When he wasn't in the wards, he was performing autopsies (the cutting open of dead bodies to examine the internal organs) to try to find out more about the germ he thought was causing the disease.

A sergeant who served with Fleming at this time described him "standing on a cold winter's morning with ice and snow everywhere . . . carrying out an autopsy on a table, while on another table another corpse lay awaiting its turn. We had six autopsies to do that morning! It was Christmas Day."

Fleming spent sleepless hours studying the germs that he found in the diseased flesh of the corpses. He found that there were several variations of *Haemophilus influenzae* and that not always the same kind was found in cases of Spanish influenza. He concluded that *Haemophilus influenzae* was probably not the primary cause of the disease, but rather the germ of a secondary infection that was masking the real cause.

He was right, but the real virus eluded him. Before the epidemic wore itself out, because the hospital was so shorthanded, Fleming—like all the staff—would himself carry the corpses to the cemetery. The devastation of the influenza epidemic affected him deeply.

"Surrounded by all those infected wounds," Fleming recalled later, "by men who were suffering and dying without our being able to do anything to help them, I was consumed by a desire to discover, after all this struggling and waiting, something which would kill those microbes."

World War I ended in November 1918, and by 1919 the influenza epidemic was waning. Fleming was discharged from the army. He returned to England, to his wife, and to his research work. His search for "something which would kill those microbes" was about to begin.

> **"I was consumed by a desire to discover, after all this struggling and waiting, something which would kill those microbes."**

THE DISCOVERY OF LYSOZYME

"When I was a young doctor in the '14-'18 war," Alexander Fleming once told a colleague, "I realized that every living thing must, *in all its parts,* have an effective defense-mechanism; otherwise, no living organism could continue to exist. The bacteria would invade and destroy it." This was the belief that would lead Fleming to discover *lysozyme,* a substance manufactured in many life-forms that defends them against germs. Lysozyme does not work on all germs, to be sure, but it is effective on so many that its discovery is now hailed by many scientists as a major accomplishment of the twentieth century. Its importance was not, however, recognized during the first years after Fleming identified it and began studying its effects.

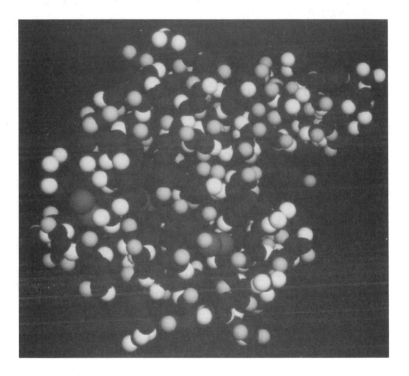

This computer-generated model shows the molecular structure of lysozyme. An enzyme found in tears and mucus, lysozyme helps protect body cells against bacterial infection by dissolving the bacteria. Fleming discovered lysozyme in 1921.

In part, this was due to his inability to present the fruits of his work effectively. His shortcomings as a speaker—he spoke softly, modestly, and without conviction—led his audiences to not take him very seriously. Soon after he returned from the war, he was invited by the Royal College of Surgeons to give a major lecture on wound infections and the failure of antiseptics to stop them. Alec was possibly the most experienced bacteriologist-surgeon to come out of the war, but his unconvincing manner left his listeners unimpressed.

Around this time, Alec had returned to the Inoculation Department at St. Mary's. He had started a private practice, but he gave it up when he was put in charge of the production of vaccines. In 1921, he and Sareen bought their country home, The Dhoon. It was here, a short time after they moved in, that Fleming caught a bad cold.

Nobody knew what caused the common cold. Fleming seized the opportunity to try to find the microbe responsible. He blew his nose, put a smear of the results on a glass slide and studied it under a microscope. After a few days he put the slide aside to work on other things.

During this period, Fleming had a new assistant, Dr. V. D. Allison. "Early on," Allison recalled, "Fleming began to tease me about my excessive tidiness in the laboratory. At the end of each day's work I cleaned my bench, put it in order for the next day and discarded tubes and culture plates for which I had no further use. He, for his part,

"This is interesting." kept his cultures . . . for two or three weeks until his bench was overcrowded with 40 or 50 cultures. . . . Discarding his cultures one evening, he examined one for some time, showed it to me and said, 'This is interesting.'"

The culture was of his nasal mucus that he had put on a slide. Large colonies of a particular type of common yellow microbe were growing at the edges of the slide. This was to be expected. There are germs in the air, and they always feed on the cultures used to make slides. What was different this time was that there were no germs near the smear of nasal mucus. Something in the nasal mucus had killed them.

Fleming took the live germs that had escaped, put them in a clear saline solution where they could be most accurately observed, transferred them to a slide, and added some new nasal mucus. In less than two minutes the germs were gone. "It was an astonishing and thrilling moment," Allison would testify, "the beginning of an investigation which occupied us for the next few years."

If, as Fleming suspected, this effect was due to some defensive substance always present in living things, then he reasoned that it should be found in other bodily fluids. He blinked, produced a few tears, and put them under a microscope. In less than five seconds the horde of microbes around the teardrops were dissolved. "I had never seen anything like it!" Fleming later recalled.

For the next five weeks Fleming and his assistant studied the effects of tears. They used their own tears, cutting lemon peels and squeezing them to make themselves cry. They persuaded visitors to the lab to produce tears for them. Lab attendants had to endure the "ordeal by lemon" and were paid three pennies per weep. The *St. Mary's Hospital Gazette* ran a cartoon showing children in the lab being beaten by one attendant while another collected their tears in a beaker labeled "Antiseptics."

The substance present in the nasal mucus and in the tears destroyed many different kinds of microbes. Some it dissolved quickly, some more slowly, some not at all. However, among those it killed were some germs that caused a variety of different infections. Fleming had identified a natural **germicide**—a substance manufactured by the body that kills bacteria.

"A HISTORIC EVENT"

In December 1921, Fleming spoke about his discovery to the Medical Research Club. As with his speech to the Royal Society of Surgeons two years earlier, his talk was greeted with indifference.

What is now regarded as "a historic event in medical history" was not recognized. Fleming's drone had overwhelmed its importance.

Alec shrugged it off and continued his researches. He tested human saliva to see if it had the same deadly effect on specific germs as tears and nasal mucus. It did. He decided to see what effect human tissue would have. He tested nail cuttings, skin scrapings, hairs. All of them worked on particular microbes.

If that was true of substances produced by the human body, why not substances produced by animals? Fleming ran tests on dog's saliva, teardrops from cows, and even from the sperm of different animals. Again, the substances tested killed particular germs. What about plant life? he wondered. He tested flowers and vegetables. All worked to some degree. Turnips worked best.

The whites of eggs from different birds seemed to work better than any other substance. The germicide was highly concentrated in egg whites, and it killed the strongest kinds of bacteria. When Fleming went fishing and caught a pike, he tested the eggs it was carrying and found they too were highly effective in killing some of the hardiest germs. The same was true of other fish eggs.

Fleming concluded that the germicide must be an enzyme. An enzyme is a chemical substance produced in living plant and animal cells. It is defined in T. Randall Lankford's *Integrated*

Science for Health Students as affecting "the speed of chemical reactions but is not itself chemically changed. . . . [Enzymes] cannot cause reactions to occur that would not normally occur." A particular enzyme will act on only one substance in the body. Its effect is to speed up the particular action taking place and to increase the energy fueling the action. Medical theory holds that "if all enzymes were removed from the body . . . vital body functions" (breathing and heartbeat) would slow down and death would occur.

Enzyme action is perhaps most easily understood in the Lock and Key Theory of Enzyme Action. The enzyme is the "key," and the substrate, the bodily substance on which the enzyme acts, is the "lock." The key must match the keyhole, and it will fit only certain specific locks. The point at which they meet, the active site, is the "keyhole." In the case of Fleming's original discovery, his nasal mucus was the key, the particular bacteria was the lock, and the culture slide was the active site.

SPREADING THE NEWS

Fleming discussed his research with his longtime boss, Sir Almroth Wright. The older man understood the importance of Fleming's work. It was agreed that Fleming should prepare another paper and that Wright would himself present it to the Royal Society. At that time the Society was, according to Fleming's biographer Gwyn Macfarlane, "the most respected, influential and exclusive scientific society in the world." However, when Wright read the paper, "On a Remarkable Bacteriolytic Element Found in Tissues and Secretions," in 1922, Fleming's work was again shrugged off as not very important. Before he made this presentation, Wright had given a name to the enzyme Fleming had discovered. He called it **lysozyme.** Loosely constructed from the Greek, *lysozyme* means an enzyme that dissolves substances—in this instance, the cell walls, or outer coverings, of the bacteria.

Despite the lack of interest shown in Fleming's work, Wright entered his name for membership in the Fellowship of the Royal Society. Acceptance would mean "one of the highest honors that a scientist could achieve." But Fleming was rejected. He continued to be denied membership for each of the next 20 years.

Undeterred, Fleming continued his work on lysozyme throughout the 1920s. He prepared solutions from egg white, targeted specific microbes, and injected rabbits. It killed germs in the animals' bloodstreams over varying periods of time. Encouraged, he injected human patients with similar solutions of lysozyme. Again the results were encouraging but limited. Only the mildest strains of bacteria were overcome by the lysozyme. Killer microbes, those that caused the deadliest diseases, were not dissolved by the lysozyme. Somehow they had adjusted over the centuries and built up a resistance to it.

LEISURE TIME

During this period, Alec's focus on microbes took a strange turn. He recognized that to work effectively he had to set aside time for recreation. It was necessary to clear the brain. At The Dhoon he gardened and played golf and croquet with Sareen. He went fishing and sometimes the two of them roamed the countryside in search of antiques. He played billiards and snooker in London, and then the day came when he took up painting.

His art was not ordinary. He had noticed that germ cultures "when grown in an incubator produce vivid colors." These were the materials Alec used to produce his abstract paintings. Somewhat to his own surprise, they were taken seriously, and he was included in an exhibition of impressionists in the London Exhibition of the Medical Arts Society. That was gratifying to Alec, but not as gratifying as the birth of his son Robert on March 17, 1924.

BACK IN THE LAB

Robert was two years old when a young eye doctor named Frederick Ridley joined Fleming's staff at St. Mary's. He and Fleming worked together to study the effects of lysozyme on the eye disease called conjunctivitis. Commonly known as red eye, this condition, which results in swelling, is caused by a microbe called staphylococcus. Lysozyme dissolved the staphylococcus. However, by treating the microbe with weak strains of lysozyme and gradually increasing their strength, it could be made resistant to the germicide. When this was done, the microbe regained its ability to cause conjunctivitis.

Fleming and Ridley had shown how germs could develop an immunity to lysozyme. This raised the possibility of destroying that immunity. But first they had to come to grips with another problem. They could squeeze out enough tears, or find enough egg white to perform lab experiments, but that was a far cry from having enough lysozyme to fight disease. A way had to be found to extract lysozyme in large enough quantities that it could be produced as a medicine. "If we had this substance pure," Fleming said, "it ought to be possible to maintain in the body a concentration which would kill certain bacteria." Along with Ridley, who had a background in organic chemistry, Fleming worked on this for two years.

They did not succeed in finding a way to extract lysozyme in quantity. It would be 1937 before this was finally done. Today lysozyme is used to fight many diseases and as a preservative in many foods. Today Fleming's name is known everywhere because of penicillin, not for lysozyme.

CHAPTER SEVEN

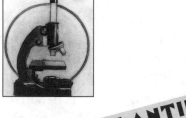

THE FIRST ANTIBIOTIC

An **antibiotic** is a substance that fights against germs that cause infection and disease. Penicillin, a substance produced by a specific mold microbe, was the first antibiotic to be discovered. Later studies showed that it was effective because it weakened the cell walls of bacteria. The cell walls eventually burst, killing the bacteria.

Throughout 1929 and the early 1930s, Fleming and various assistants and chemists experimented with many different ways to extract pure penicillin from its broth. Their aim was to obtain it in the form of crystals. Once this was done, it could be made suitable for injection. It could be mass produced. However, while they thought they were on the brink of solving the problem several times, they always met with disappointment.

During this time, work had begun to renovate and add to the buildings housing St. Mary's. It was completed in December 1933, and the king and queen of England attended the official reopening. Sir Almroth Wright had asked his staff, including Fleming, to arrange demonstrations of their work for the royal couple. Fleming hung a group of his germ paintings created with colored bacteria arranged in patterns. These included landscapes, ballerinas, and a red-white-and-blue British flag. The queen looked at the paintings, blinked, and remarked, "Yes—but what *good* is it?"

Some good. On June 14, 1931, it was determined that penicillin stopped the growth of microbes causing gangrene. Important as this was, it received only slight notice in the scientific world. Fleming doggedly kept on working, developing both a penicillin liquid and an ointment. The liquid cleared up infants' eye infections, and the ointment was effective on surface wounds. Writing in the *Journal of Pathology* in 1932, he reported that penicillin was "superior to dressings containing potent chemicals."

The problem was still how to extract and purify penicillin for mass use, but Fleming was sure that would be solved. Meanwhile, he sent culture dishes of penicillin to laboratories and hospitals in England and abroad so that other bacteriologists might grow it and test it on various germs. It would take time before penicillin would receive the recognition it deserved, but its germ-fighting possibilities were now at least beginning to be discussed in medical circles around the world.

GOOD MEDICINE

Fleming himself could not continue to give his discovery as much attention as he might have

This electron micrograph shows the fruiting bodies of penicillin growing on a moldy piece of cheddar cheese. After Fleming's discovery of penicillin, the next challenge was to extract and purify the antibiotic for mass use.

liked. He was in charge of the production of vaccines at St. Mary's. This meant improving methods of isolating those microbes that caused specific diseases and improving the methods of manufacturing weak strains of them for vaccines. He successfully produced vaccines for certain kinds of influenza, whooping cough, and acne.

His work on vaccines led him to coauthor a book on the subject with G. F. Petrie of the Lister Institute. Titled *Recent Advances in Serum and Vaccine Therapy,* it was published in 1934. Recognition of Fleming's work with vaccines led to an extended trip throughout the Middle East and parts of Europe in 1935. That same year, Howard Florey and Ernst Chain came to work at Oxford University.

The 37-year-old Australian Florey was appointed Professor of Pathology at the world-renowned university. He hired Chain, who was 29 years old, to run his Department of Biochemistry. Both Florey and Chain had experimented with lysozyme. Both were admirers of Fleming's work.

Soon after he came to Oxford, Chain ran across Fleming's 1929 paper on penicillin. Already recognized as a brilliant chemist, the German-Jewish refugee understood and appreciated the problem of extracting pure penicillin without making it unstable. Like Fleming, he became convinced that there must be a way to do it.

Pathologist Howard Florey (1898–68) assembled the team at Oxford University that produced penicillin in large enough quantities to conduct clinical studies. The team's research proved that penicillin was effective against bacterial infections.

But the necessary research would take money. Chain discussed this with Florey, who immediately saw the possibilities. He became as enthusiastic as Chain, and he persuaded the Rockefeller Foundation to give them a $5,000 grant to work on penicillin.

At this time Fleming was being drawn into experiments on a new disease-fighting substance. It was called *Protonsil,* and it was the creation of German doctor-chemist Gerhard Domagk. A chemical rather than a vaccine, Protonsil was a red dye that killed germs. It was the first of a number of synthetic chemicals that would be called **sulfonamides,** or sulfa drugs.

Sulfa drugs would work on such diseases as pneumonia, meningitis, and scarlet fever, but they would be useless against many other kinds of disease-causing and infection-spreading bacteria. They did not work well where pus was present, and their side effects included skin rashes and frequent vomiting. Most dangerously, they some-times interfered with the body's production of phagocytes. This lowered a patient's resistance to infection and could be fatal.

Fleming tested the new drugs on a variety of bacteria growing in culture dishes. He noted that they did not actually kill the microbes. What they did was interfere with their growing and spreading. How then did they cure disease in the body?

Fleming concluded that the microbes were so weakened by the sulfa drugs that they became easy prey to the phagocytes. If that was so, Fleming decided, then the sulfa drugs could be used most effectively in combination with vaccines, which would strengthen phagocytes to fight diseases.

A DEVASTATING LOSS

Fleming experimented to determine practical combinations of sulfa drugs and vaccines over the next few years. Sadly, the successes he had did not come in time to prevent a family tragedy. In 1937, his brother John came down with pneumonia.

This was John's second attack. Two years earlier he had recovered from pneumonia thanks to a recently developed serum. This time, however, the serum had no effect. In another decade, penicillin would have destroyed the pneumonia germs. But in 1937 it had not yet been extracted in a pure enough form to treat John's pneumonia. Nor was there yet a sulfa drug that would fight the particular form of the disease John had.

John died. Not being able to save him was a hard blow to Alec. John's wife Elizabeth, the sister of Alec's wife, Sareen, also took it very hard. When she sank into a deep depression, Alec and Sareen took her to live with them at The Dhoon. She remained a gloomy presence in the house after Alec and Sareen's son Robert went off to boarding school.

PERFECTING PENICILLIN

Alec went back to his experiments on combinations of sulfa drugs and vaccines. He continued with this work for the next two years. Meanwhile, at Oxford, Florey and Chain were devoting more and more time to their efforts to find a way to purify penicillin.

They found that when they tried to extract penicillin in a concentrated form by evaporating it, the penicillin's instability became worse. A new process was tried in which the penicillin juice was put in a vacuum chamber and made into a gas. They were left with a yellow-brown powder. This was a more pure form of penicillin than any that had been extracted before, but it still had too many impurities to be used in treating anything but the surface areas of people.

Florey and Chain melted the powder in methanol, a simple type of alcohol. This got rid of most of the remaining impurities. Now they tested this penicillin powder for stability. They gave it to mice and there were no ill effects. Further tests with microbe cultures showed that the powder was a thousand times stronger than raw penicillin. With many bacteria, it was also far more effective than the latest sulfa drugs.

WAR AGAIN

It was during this period that World War II broke out. Nazi Germany invaded Poland, and when France and England declared war Western

Ernst Chain (1906–1979) examined the biochemistry of Fleming's two great discoveries, lysozyme and penicillin.

Europe became the battlefield. Nazi troops quickly swept over Holland, Belgium, and France.

Fleming had been in the United States on September 3, 1939, when England declared war on Germany. He had come to read a paper at the Third International Congress of Microbiology in New York. The subject of the paper was the effects of combining vaccines with chemicals, such as sulfa drugs.

Sareen was with him, and the couple had intended to see something of the United States during their visit. They dropped these plans when war broke out. Instead they took the first boat home so that Fleming, a leading expert on the treatment of war wounds, might offer his services to his country.

The war was going very badly for Great Britain in May 1940 when Florey and Chain deliberately injected three different types of deadly microbes into three groups of mice. In each group some of the mice were given penicillin and some were not. Those who received it got better. All of the other mice died.

On May 29, 1940, the 340,000-man British Army, along with other Allied troops, retreated across the channel to England to avoid being captured by the Germans. The Nazis were now in control of almost all of Europe. There was great fear that they would soon invade England.

In September 1941, one of London's double-decker buses lumbers through the wreckage caused by German bombing. The daily ravages of the war pushed Fleming, Florey, and Chain to find a way to produce large quantities of penicillin.

The war casualties had been high, and Florey and Chain knew that if penicillin could only be made available it would save countless lives. On July 1, they infected 50 mice with streptococci. Half of the mice were treated with penicillin on a regular basis for 48 hours. The rest received no treatment. After 16 hours all of the untreated mice were dead. Of the 25 others, 24 completely recovered.

With the heavy bombing of Great Britain beginning, Florey and Chain published their findings in *Lancet*. They reported that penicillin acted particularly well against the specific microbes that caused gangrene. Alexander Fleming, who was working on war wounds for the government, was particularly interested in fighting gangrene. He was gratified to read that his discovery, penicillin, might now be made available in a form which would do this. The problem of extracting and purifying penicillin without losing its ability to fight microbes had been solved!

Fleming traveled to Oxford on September 2, 1940. "I've come to see what you've been doing with my old penicillin," was his greeting to Florey and Chain. They were astonished by his visit.

"Boo!"

"They thought I was dead, you see," Fleming explained later. "The late Professor Fleming." He was asked what he said to them then. His one word answer was typical of his dry wit: "Boo!" That's what he said. Just "Boo!"

CHAPTER EIGHT

LOVE AND DEATH

The men who made treatment with penicillin possible—Fleming, Florey, and Chain—worked feverishly through the first half of 1941 to arrange for the antibiotic to be manufactured in large quantities. With the war raging, they counted the days in lives lost, lives that could have been saved by penicillin. Finally they accepted the fact that war-torn England simply did not have the resources to produce the new drug.

They turned to the United States, which in June 1941 was not yet at war. There, with the backing of the Rockefeller Foundation, a manufacturer was found. The drug was produced, tested, and used in larger and larger quantities. As its reputation grew, the demand for it skyrocketed. After December 7, 1941, when the Japanese attacked the U.S. naval base at Pearl Harbor, Hawaii, and the United States entered the war, production of penicillin became a national priority. It was shipped to Allied fighting units, including England's. Civilians would not have access to the new miracle drug until after the war.

A major wartime producer of penicillin was the North Regional Laboratory in Peoria, Illinois. The company was trying to develop a new strain of penicillin that would grow at a faster rate. One of the lab's employees, Mary Hunt, was assigned to shop for rotting fruits and vegetables that might have more fast-acting molds than the ones being used. Fellow employees nicknamed her Moldy Mary. The name paid off when she turned up a cantaloupe with a mold that produced a strain of penicillin that grew 200 times as quickly as the penicillin mold Fleming had discovered.

Already penicillin was being hailed as a miracle drug. To some extent, Fleming agreed. "It is a miracle," he granted, "and it will save lives by thousands." But he added that he had never claimed that penicillin could cure all diseases. He thought people should understand that penicillin acted on some microbes but not on others.

"It is a miracle, and it will save lives by thousands."

Fleming was also modest about taking credit for his work. "Nature makes penicillin," he protested. "I just found it." Nevertheless, he was honored. After twenty years of being rejected, Alexander Fleming was made a member by the Royal Society. Then in 1943, along with Howard Florey, he was knighted for his discovery of penicillin. With German bombs falling on London, the ceremony was held in the air raid shelter basement of Buckingham Palace.

The air raids were no joke. In 1944 the Germans were sending over V-2 rockets—guided missiles filled with incendiary explosives—on a daily basis. On two occasions Fleming narrowly escaped the explosions that damaged his London apartment. "When I saw the entire window-frame moving towards me," he remarked after the second episode, "I decided to get out of bed."

"Nature makes penicillin. I just found it."

On June 6, 1944, the D-Day landings—in which United States, English, and French troops invaded Europe—took place. Casualties were high and penicillin made a real difference between life and death. Of the wounded soldiers who were treated with penicillin, 95 percent recovered. This accomplishment inspired the construction of the world's largest penicillin factory in England in November 1944.

Six months later, on May 7, 1945, Germany surrendered, and the war in Europe was over.

That summer the manufacturers of penicillin in the United States invited Fleming to a banquet in his honor at the Waldorf Astoria in New York. In recognition of penicillin's benefit to humanity, they presented him with a check for $100,000. Fleming turned the money over to St. Mary's to be used for research.

During this trip Fleming visited the Pfizer laboratories in Brooklyn, New York, where research on penicillin was going forward. He was dazzled by the lab's antiseptic appearance, the scrubbed counter surfaces, and the gleaming instruments. "If I had been working in these conditions," he chuckled, "I should never have found penicillin."

> **"If I had been working in these conditions, I should never have found penicillin."**

On his return to England, Alec was notified that he, Sir Howard Florey, and Professor Ernst Chain had won the Nobel Prize for Medicine, the highest honor the world had to offer. Then Sir Almroth Wright retired, and Fleming became the head of the Institute of Microbiology at St. Mary's. He did not much care, however, for being an administrator and stole time to work on his research whenever he could. He was helped in this by a new member of the department, a 34-year-old woman bacteriologist from Greece named Amalia Voureka.

Fleming (second from left) joins other Nobel laureates in Stockholm, Sweden. He shared the 1945 Nobel Prize for physiology and medicine with Ernst Chain (center) and Howard Florey (far right).

A NEW PARTNER

Voureka was a remarkable woman. During World War II, she and her husband had been Greek resistance fighters. Both had been caught and jailed. They had survived, but their marriage had not. Although they had not divorced, they had gone their separate ways.

Amalia had been a medical student before the war. Now she completed her studies, specializing in bacteriology. She was much impressed by penicillin and deliberately went to London to apply for a job that had opened up in Fleming's department. She so impressed those who interviewed her, including Fleming, that she was hired over some very stiff competition. She was the first female to work in a department that, until recently, had been run by the woman-hating Sir Almroth Wright. Her abilities were such that before long she had become indispensable to Fleming.

This was a sad time for him. On April 30, 1947, at the age of 87, Wright died. He had been a difficult man of many prejudices, but he had always been an inspiration to Alec. Wright's genius had challenged Fleming always to do his best work.

The following year, on a trip to Spain, Sareen collapsed. Back in London her condition grew worse. She may have had Parkinson's disease, but that diagnosis was never confirmed. She was moved to a nursing home where her situation became grave. Alec despaired. "The most horrible thing about it is that penicillin can do nothing for her," he realized. "When John died it had not been perfected: now it has, but it is useless in Sareen's case."

Sareen lingered for over a year. Then, on October 28, 1949, she died. They had been married for 34 years. "My life is broken," Alec grieved.

Work gave him relief. He plunged into studies of the effects of penicillin on different microbes. Amalia Voureka, now his chief assistant, worked alongside him tirelessly.

Together, they wrote and published a paper on this work in the *Journal of General Microbiology*. Voureka, who spoke several languages, translated many of Fleming's writings. She took trips abroad with him, acting as his translator. Back in England, he invited her to The Dhoon, and soon they were spending long weekends there, bringing work with them.

During one of these weekends Alec asked Amalia about the state of her marriage. She and her husband had been separated for some 15 years. She told him that they had agreed on a divorce and that it would soon be final.

Not long afterwards they traveled to Athens, where Fleming received an award and Amalia translated his acceptance speech into her native tongue. Alec was 71 years old and had been a widower for three years. Amalia was 39. One night in Athens Alec turned to Amalia and said something in a low tone of voice that was slurred by a Scotch burr made thicker by nervousness. "Did you say anything?" Amalia asked. "I asked you to marry me," was the reply. Amalia blinked. It took a moment for his meaning to sink in. Then she answered him. She said "yes."

Before they could be married, Alec was off to India as a member of the World Health Delegation. Amalia remained in Athens. It would be April 1953 before they would meet again in London with the wedding set for the ninth of that month. Alec's colleagues and friends admired Amalia. They commented on her admirable character and her kindness, sincerity, and bravery. In her work she had a reputation for discipline and intelligence. They were happy together and traveled widely. Co-workers at St. Mary's remarked on the change in Alec. Where once he had seemed reserved, almost stern, now he wore a radiant smile.

THE LAST DAY

On March 11, 1955, Alec and Amalia were in London. Alec woke up looking forward to the day. He and Amalia were to dine that evening with Eleanor Roosevelt and the film actor Douglas Fairbanks Jr. Alec started the day with a bath. When he came out of the bathroom, Amalia noticed that he looked pale. Alec told her he felt nauseous. Alarmed, she called the doctor, who was busy with other patients.

Alec said he felt better. Amalia left him to get dressed. The doctor called back and Alec took the call. "Is it urgent?" the doctor asked. "No urgency whatever," Alec told him. "Look after your other patients first."

Fleming and Amalia Voureku chat with well-wishers outside a Greek Orthodox Church after their 1953 wedding ceremony.

Amalia returned, and Alec asked her to comb his hair. As she was doing this, she realized he was covered with cold sweat. He complained of a pain in his chest. She worried that it might be his heart.

"It's not the heart," he reassured her. A moment later his head fell forward. Sir Alexander Fleming, the discoverer of penicillin, was dead.

AFTERWORD

The direction in which Fleming's discovery of penicillin pointed for those who followed him is more important than the discovery itself. He had found, in a common mold, a unique group of microbes. They produced a substance that interfered with the growth of other bacteria that caused serious diseases and infections. This began the search for other molds with other microbes that produced substances to fight disease-causing germs.

Penicillin, as powerful a weapon as it has proved to be in the war against disease, was the first of many antibiotics to be developed by researchers. Those that have followed have changed the practice of medicine. They have saved countless lives that might have been lost without them. Among the many diseases that were once fatal but are now successfully overcome by penicillin and other antibiotics are pneumonia, rheumatic fever, spinal meningitis, typhoid fever, syphilis, gonorrhea, and gangrene. Experiments are underway to determine the effectiveness of newly developed antibiotics in various types of cancer. So far such viruses as HIV (human immune deficiency virus), the AIDS virus, have been resistant to antibiotics, but the research continues.

The possibilities are unlimited. Alexander Fleming recognized this. "The extraordinary merit of penicillin has trained the searchlight on a new field," he pointed out in 1943 when its infection-fighting potential was first recognized. He went on to predict that "diseases now untouchable will be conquered." This has already happened, and indications are that it will continue to happen. And each new breakthrough, each new disease overcome, will be another tribute to the scientist Alexander Fleming who looked at a mold and saw a miracle in the making.

Fleming had this advice for young people attracted to medicine and research: "Never neglect an extraordinary appearance or happening. . . . But I warn you of the danger of first sitting and waiting till chance offers something. . . . Work hard, work well, do not clutter up the mind too much with precedents, and be prepared to accept such good fortune as the gods offer."

A NOTE ON SOURCES

I used many materials in writing this biography, but the following are the most important. The two major Fleming biographies are Gwyn Mac-farlane's *Alexander Fleming: The Man and the Myth* (Cambridge, MA.: Harvard University Press, 1984) and Andre Maurois's *The Life of Sir Alexander Fleming: Discoverer of Penicillin* (New York: Dutton, 1959). Steven Otfinoski's *Alexander Fleming: Conquering Disease with Penicillin* (New York: Facts On File, 1992), written for young adults, provides an excellent account of Fleming's life. I also consulted L. J. Ludovici's *Fleming: Discoverer of Penicillin* (London: Andrew Dakers, 1952) and W. A. C. Bullock's *The Man Who Discovered Penicillin: The Life of Sir Alexander Fleming* (London: Faber and Faber, 1963). A brief profile of Fleming written during his lifetime can be found in the 1944 edition of *Current Biography*

(New York: Wilson, 1944). Most of the quotations in this book may be found in these sources.

To gain a firmer grasp of the scientific topics discussed in this book, I referred to Paul De Kruif's *Microbe Hunters* (New York: Harcourt Brace, 1926), *The Merck Manual,* Vol. I (Rahway, NJ: Merck & Co., 1987), and T. Randall Lankford's *Integrated Science for Health Students: Third Edition* (Reston, VA: Reston, 1984). For detailed information on penicillin and its discovery, Francine Jacobs's *Breakthrough: The True Story of Penicillin* (New York: Putnam, 1985) and John C. Sheehan's *The Enchanted Ring: The Untold Story of Penicillin* (Cambridge, MA: MIT Press, 1982) were very helpful sources.

FOR FURTHER READING

De Kruif, Paul. *Microbe Hunters*. New York: Harcourt Brace, 1926.

LaVert, Marianne. *AIDS: A Handbook for the Future*. Brookfield, CT: Millbrook, 1996.

Murphy, Pat, Ellen Klages, and the staff of the Exploratorium. *The Science Explorer*. New York: Henry Holt/Owl Books, 1996.

Nourse, Alan E. *The Virus Invaders*. New York: Franklin Watts, 1992.

Otfinoski, Steven. *Alexander Fleming: Conquering Disease With Penicillin*. New York: Facts On File, 1992.

Poynter, Margaret. *Marie Curie: Discoverer of Radium*. Springfield, NJ: Enslow, 1994.

Rainis, Kenneth G. and Bruce J. Russell. *Guide to Microlife*. Danbury, CT: Franklin Watts, 1996.

Stwertka, Eve, and Albert Stwertka. *Microscope: How to Use and Enjoy It*. Columbus, OH: Silver Burdett Press, 1988.

GLOSSARY

anaerobes germs that die when exposed to the air

antibiotic a natural substance produced by certain microbes that destroys, or stops the growth of, bacteria

antiseptic a chemical that fights infection on living tissue

atoxyls arsenic-based drugs developed to kill syphilis germs; the most successful was salvarsan

bacteria single-celled plantlike microorganisms that can enter the body and cause infection

bacteriologist a scientist who studies bacteria with the aim of fighting diseases caused by them

culture the growing of bacteria in a specially prepared substance that will nourish them so that their activity may be scientifically studied

enzyme an organic substance produced by plant and animal cells that affects the rate at which chemical reactions occur in them; an enzyme speeds up or slows down the processes by which those cells grow or multiply but neither takes part in the reaction nor is used up by them

epidemic the rapid spread of disease among large numbers of people

gangrene decay that takes place on living tissue when the blood supply is blocked by injury or disease

germicide anything that kills germs

germs tiny organisms, such as bacteria or microbes; many germs cause diseases, but some germs fight them

immune system a collection of organs and tissues that work together to help the body defend itself against disease

immunity resistance to disease-causing bacteria

infection the invasion of the body by microbes that grow there and spread; a condition characterized by redness, swelling, and heat

lysozyme substance produced by many life-forms which defends them against some germs

microbes disease-causing bacteria

mold a fuzzy growth that thrives on decay in warm, moist places

nucleic acid a substance that bears genetic information; DNA and RNA are the two types of nucleic acid

parasite an organism that lives on or in another organism and feeds on it

penicillin an antibiotic that is produced by certain molds

phagocyte a kind of white blood cell that engulfs bacteria

salvarsan the chemical that kills syphilis bacteria

serum a fluid containing a weak strain of bacteria that is injected into a person to create immunity to diseases caused by the full-strength bacteria; also, the fluid portion of the blood

Staphylococcus a family of bacteria that causes pus-producing infections

Streptococcus a family of bacteria that causes tonsillitis, arthritis, scarlet fever, and other diseases

sulfonamides synthetic chemicals that kill certain bacteria; also known as sulfa drugs

syphilis a disease transferred by sexual contact or from mother to child at birth

tetanus an infectious wound disease that can be fatal; also known as lockjaw

vaccine a solution of dead bacteria injected to increase immunity to a specific disease

virus a weakened or nonliving infectious particle containing nucleic acids surrounded by a protein coat; about one-tenth the size of most microbes, it must get inside of living cells to become active and reproduce; viruses cause polio, smallpox, rabies, and the common cold

white blood cells colorless blood cells that are a key component of the human immune system

CHRONOLOGY

1881	Alexander Fleming is born on August 6.
1893	Enters Kilmarnock Academy.
1895	Enrolls in the Polytechnic Institute in London.
1897	Begins working for the American Line.
1900	Enlists in the London Scottish Rifle Volunteers.
1901	Enters St. Mary's Hospital Medical School.
1906	Graduates as a doctor from the Royal College of Surgeons; joins Sir Almroth Wright's Inoculation Department at St. Mary's.
1908	Becomes Casualty House surgeon at St. Mary's.
1909	Publishes research on acne in *Lancet*.
1909–12	Gains notoriety treating syphilus patients with salvarsan.

1914	Becomes a lieutenant in the British Army Medical Services.
1915	Marries Sareen McElroy on December 23.
1918	Tries in vain to isolate in influenza virus.
1922	Reveals discovery of lysozyme.
1928	Stumbles on the effect of the pencillin mold.
1929	Successfully treats a sinus infection with pencillin.
1931	Pencillin stops growth of microbes causing gangrene.
1934	Fleming publishes major work on serum and vaccine therapy.
1939	Howard Florey and Ernst Chain develop method to extract pure pencillin.
1941	Rockefeller Foundation backs mass production of pencillin in United States.
1945	Fleming, Florey, and Chain are jointly awarded the Nobel Prize.
1949	Sareen Fleming dies.
1953	Alexander Fleming marries Amalia Voureka.
1955	Dies on March 11 at age 73.

INDEX

Page numbers in *italics* indicate illustrations.

ABOUT THE AUTHOR

Ted Gottfried is the author of more than 50 books, both fiction and nonfiction, including biographies of Alan Turing, Enrico Fermi, and James Baldwin. Mr. Gottfried has taught writing at New York University, Baruch College, and other institutions. He is married and lives in New York City.